Civil Engineering Formulas and Equations for Students

100 Essential Equations and Formulas for Scientists and Engineers

Part III: Water Resources and Environmental Engineering

Peter I. Kattan

Petra Books
www.petrabooks.com

Civil Engineering Formulas and Equations for Students

100 Essential Equations and Formulas for Scientists and Engineers

Part III: Water Resources and

Environmental Engineering

To contact the author send email to:
pkattan@petrabooks.com

Civil Engineering Formulas and Equations for Students

100 Essential Equations and Formulas for Scientists and Engineers

Part III: Water Resources and

Environmental Engineering

ISBN: 979-8-3493-3135-0

Contents

Preface

This is a little book for students who wish to have the essential formulas and equations of water resources and environmental engineering in a single easily accessible source. In about 40 pages, the 100 most essential civil engineering formulas are listed. Unlike other large books on this topic, there is no need to go through hundreds of pages and thousands of formulas for the student to get the basic equations. The author has searched several books on engineering formulas and tables and selected only those equations which are essential to the student. The civil engineering formulas and equations listed in this book are useful for students and researchers in various fields including environmental engineering and water resouces, etc. Only the most elementary and basic topics are covered including formulas for fluid mechanics, fluid pressure, fundamentals of fluid flow, water resources modeling including mathematical models, pipe flow, laminar and turbulent flow, minor losses in pipes, orifices, fluid jets, pipe stresses, temperature expansion in pipes, open channel flow, hydraulic jump, groundwater, flow from wells, environmental engineering, sewer design, velocity formulas, storm-water inlets, side weirs, friction loss in full pipe flow, work and pump efficiency, characteristics of domestic wastewater, BOD, CBOD, and COD. This is the third book in this series and is dedicated to water resources and environmental engineering. Other volumes area available in other areas of civil engineering.

Civil Engineering Formulas and Equations for Students

100 Essential Equations and Formulas for Scientists and Engineers

Part III: Water Resources and

Environmental Engineering

Fluid mechanics

Properties of fluids

Capilarity

$$h = \frac{2\sigma\cos\theta}{(w_1 - w_2)r}$$

h = capillary rise, ft
σ = surface tension, lb/ft
w_1 and w_2 = specific weights of fluids below and above meniscus, respectively, lb/ft
θ = angle of contact
r = radius of capillary tube, ft

Viscocity

Viscosity, μ of a fluid, also called the coefficient of viscosity, absolute viscosity, or dynamic viscosity, is a measure of its resistance to flow.

$$\mu = \frac{\tau}{dV/dy}$$

τ = shearing stress, lb/ft^2

V = velocity , ft/s
y = depth, ft

Fluid pressure

p, the pressure, psf, at depth h.

$$p = wh \ , \ \ h = \frac{p}{w}$$

The above equation gives the depth of water h of specific weight w required to produce a gage pressure p. By adding atmospheric pressure p_a to the equation, absolute pressure p_{ab} is obtained:

$$p_{ab} = wh + p_a$$

Submerged and floating bodies

A floating body displaces a volume of liquid equal to its weight. The buoyant force acts vertically through the center of buoyancy c.b., which is located at the center of gravity of the volume of liquid displaced

Fundamentals of fluid flow

For fluid energy, the law of conservation of energy is represented by the Bernoulli equation:

$$Z_1 + \frac{p_1}{w} + \frac{V_1^2}{2g} = Z_2 + \frac{p_2}{w} + \frac{V_2^2}{2g}$$

Z_1 = elevation, ft , at any point 1 of flowing fluid above an arbitrary datum
Z_2 = elevation, ft , at downstream point in fluid above same datum

$$\begin{aligned} p_1 &= \text{ pressure at 1, psf} \\ p_2 &= \text{ pressure at 2, psf} \\ w &= \text{ specific weight of fluid, lb/ft}^3 \\ V_1 &= \text{ velocity of fluid at 1, ft/s} \\ V_2 &= \text{ velocity of fluid at 2, ft/s} \\ g &= \text{ acceleration due to gravity, 32.2ft/s}^2 \end{aligned}$$

The left side of the equation sums the total energy per unit weight of fluid at 1 , and the right side, the total energy per unit weight at 2.

This term h_f, when added to the downstream side of the equation, yields the form of the Bernoulli equation most frequently used:

$$Z_1+\frac{p_1}{w}+\frac{V_1^2}{2g}=Z_2+\frac{p_2}{w}+\frac{V_2^2}{2g}+h_f$$

The total energy, in the elemental volume of fluid is

$$E=WZ_a+\frac{Wp_a}{w}+\frac{WV_a^2}{2g}$$

Dividing both sides of the equation by W yields the energy per unit weight of flowing fluid, or the total head ft :

$$H=Z_a+\frac{p_a}{w}+\frac{V_a^2}{2g}$$

p_a/w is called pressure head; $V_a^2/2g$, velocity head.

The momentum equation, or impulse-momentum equation is given as follows:

$$F_x=pQ\Delta V_x$$

F_x = summation of all forces in X direction per unit time causing change in momentum in X direction, lb
ρ = density of flowing fluid, lb $\cdot$ s^2/ft^4 (specific weight divided by g)

Q = flow rate, ft^3/s

ΔV_x = change in velocity in X direction, ft/s

Water resources modeling

Mathematical models

The Froude number is

$$\mathbf{F} = \frac{V}{\sqrt{Lg}}$$

$\mathbf{F}$ = Froude number (dimensionless)
V = velocity of fluid, ft/s
L = linear dimension (characteristic, such as depth or diameter), ft
g = acceleration due to gravity, 32.2ft/s^2

For hydraulic structures, such as spillways and weirs, where there is a rapidly changing watersurface profile, the two predominant forces are inertia and gravity. Therefore, the Froude numbers of the model and prototype are equated:

$$\mathbf{F}_\mathrm{m} = \mathbf{F}_\mathrm{p} \frac{V_m}{\sqrt{L_m g}} = \frac{V_p}{\sqrt{L_p g}}$$

where subscript m applies to the model and p to the prototype.

Squaring both sides of the above equation and grouping like terms yields

$$\frac{V_m^2}{V_p^2} = \frac{L_m}{L_p}$$

Let $V_r = V_m/V_p$ and $L_r = L_m/L_p$. Then

$$V_r^2 = L_r \; V_r = L_r^{1/2}$$

The subscript r indicates ratio of quantity in model to that in prototype.

The discharge ratio is determined as follows:

$$Q_r = V_r A_r$$

Since $V_r = L_r^{1/2}$ and A_r = area ratio = L_r^2,

$$Q_r = V_r A_r = L_r^{5/2}$$

The Reynolds number is

$$\mathbf{R} = \frac{VL}{v}$$

R is dimensionless, and v is the kinematic viscosity of fluid, ft^2/s.

The following relations are obtained by equating Reynolds numbers of the model and prototype:

$$\frac{V_m L_m}{v_m} = \frac{V_p L_p}{v_p}$$

$$V_r = \frac{v_r}{L_r}$$

The Weber number is

$$\mathbf{W} = \frac{V^2 L \rho}{\sigma}$$

ρ = density of fluid, $\mathrm{lb} \cdot \mathrm{s}^2/\mathrm{ft}^4$ (specific weight divided by g)
σ = surface tension of fluid, psf

The velocity ratio is determined as follows:

$$\frac{V_m^2 L_m \rho_m}{\sigma_m} = \frac{V_p^2 L_p \rho_p}{\sigma_p}$$

$$V_r^2 = \frac{\sigma_r}{\rho_r L_r}$$

Pipe flow

Laminar flow

A dimensionless parameter called the Reynolds number has been found to be a reliable criterion for the determination of laminar or turbulent flow. It is the ratio of inertial forces to viscous forces, and is given by

$$\mathbf{R} = \frac{VD\rho}{\mu} = \frac{VD}{v}$$

V = fluid velocity, ft/s
D = pipe diameter, ft
ρ = density of fluid, $\mathrm{lb} \cdot \mathrm{s}^2/\mathrm{ft}^4$ (specific weight divided by $g, 32.2\mathrm{ft/s}^2$)
μ = viscosity of fluid $\mathrm{lb} \cdot \mathrm{s/ft}^2$
$v = \mu/\rho$ = kinematic viscosity, ft^2/s

For a Reynolds number less than 2000, flow is laminar in circular pipes. When the Reynolds number is greater than 2000, laminar flow is unstable; a disturbance will probably be magnified, causing the flow to become turbulent.

In laminar flow, the following equation for head loss due to friction can be developed by considering the forces acting on a cylinder of fluid in a pipe:

$$h_f = \frac{32\mu LV}{D^2\rho g} = \frac{32\mu LV}{D^2 w}$$

h_f = head loss due to friction, ft
L = length of pipe section considered, ft
g = acceleration due to gravity, $32.2\mathrm{ft/s^2}$
w = specific weight of fluid, $\mathrm{lb/ft^3}$

Substitution of the Reynolds number yields

$$h_f = \frac{64}{\mathbf{R}}\frac{L}{D}\frac{V^2}{2g}$$

Turbulent flow

Darcy-Weisbach Formula

$$h_f = f\frac{L}{D}\frac{V^2}{2g}$$

h_f = head loss due to friction, ft
f = friction factor (see Fig. 21.19)
L = length of pipe, ft
D = diameter of pipe, ft
V = velocity of fluid, ft/s
g = acceleration due to gravity, $32.2\mathrm{ft/s^2}$

Roughness values ε (ft) for use with the Moody diagram to determine the Darcy-Weisbach friction factor f are listed in the table below.

	ε, ft
Steel pipe:	
Severe tuberculation and incrustation	0.03-0.008
General tuberculation	0.008-0.003
Heavy brush-coat asphalts, enamels, and tars	0.0030.001
Light rust	0.001-0.0005
New smooth pipe, centrifugally applied enamels	0.00020.00003

Chezy Formula

This equation holds for head loss in conduits and gives reasonably good results for high Reynolds numbers:

$$V = C\sqrt{RS}$$

V = velocity, ft/s
C = coefficient, dependent on surface roughness of conduit
S = slope of energy grade line or head loss due to friction, ft/ft of conduit
R = hydraulic radius, ft

Manning's Formula

Through experimentation, Manning concluded that the C in the Chezy equation should vary as $R^{1/6}$

$$C = \frac{1.486R^{1/6}}{n}$$

n = coefficient, dependent on surface roughness.

$$V = \frac{1.486}{n}R^{2/3}S^{1/2}$$

Upon substitution of $D/4$, where D is the pipe diameter, for the hydraulic radius of the pipe, the following equations are obtained for pipes flowing full:

$$V = \frac{0.590}{n}D^{2/3}S^{1/2}$$

$$Q = \frac{0.463}{n}D^{8/3}S^{1/2}$$

$$h_f = 4.66n^2\frac{LQ^2}{D^{16/3}}$$

$$D = \left(\frac{2.159Qn}{S^{1/2}}\right)^{3/8}$$

Q = flow, ft^3/s.

Hazen-Williams Formula

This is one of the most widely used formulas for pipe flow computations of water utilities, although it was developed for both open channels and pipe flow:

$$V = 1.318C_1R^{0.63}S^{0.54}$$

For pipes flowing full:

$$
\begin{aligned}
V &= 0.55 C_1 D^{0.63} S^{0.54} \\
Q &= 0.432 C_1 D^{2.63} S^{0.54} \\
h_f &= \frac{4.727}{D^{4.87}} L \left(\frac{Q}{C_1}\right)^{1.85} \\
D &= \frac{1.376}{S^{0.205}} \left(\frac{Q}{C_1}\right)^{0.38}
\end{aligned}
$$

where V = velocity, ft/s
C_1 = coefficient, dependent on surface roughness
R = hydraulic radius, ft
S = head loss due to friction, ft/ft of pipe
D = diameter of pipe, ft
L = length of pipe, ft
Q = discharge, ft^3/s
h_f = friction loss, ft

Values of n for Pipes, to Be Used with the Manning Formula are given in the table below.

Material of Pipe	Variation		Use in designing	
	From	To	From	To
Clean cast iron	0.011	0.015	0.013	0.015
Dirty or tuberculated cast iron	0.015	0.035		
Riveted steel or spiral steel	0.013	0.017	0.015	0.017
Welded steel	0.010	0.013	0.012	0.013
Galvanized iron	0.012	0.017	0.015	0.017

The C_1 terms in the table are in the foot-poundsecond system.

Type of pipe	C_1
Cast iron:	
New	All sizes, 130
5 years old	All sizes up to 24in, 120 24 in and over, 115
10 years old	12 in, 110 4 in, 105 30 in and over, 85
40 years old	16 in, 80 4 in, 65
Welded steel	Values the same as for cast-iron pipe, 5 years older
Riveted steel	Values the same as for cast-iron pipe, 10 years older
Wood stave	Average value, regardless of age, 120
Concrete or concretelined	Large sizes, good workmanship, steel forms, 140
	Large sizes, good workmanship, wood forms, 120

Minor losses in pipes

Sudden Enlargements

The following equation for the head loss, ft , across a sudden enlargement of pipe diameter has been determined analytically and agrees well with experimental results:

$$h_L = \frac{(V_1 - V_2)^2}{2g}$$

V_1 = velocity before enlargement, ft/s

V_2 = velocity after enlargement, ft/s

$g = 32.2 \text{ft/s}^2$

Another equation for the head loss caused by sudden enlargements was determined experimentally by Archer.

$$h_L = \frac{1.1(V_1 - V_2)^{1.92}}{2g}$$

Gradual Enlargements

The equation for the head loss due to a gradual conical enlargement of a pipe takes the following form:

$$h_L = \frac{K(V_1 - V_2)^2}{2g}$$

K = loss coefficient

Sudden Contraction

The following equation for the head loss across a sudden contraction of a pipe was determined:

$$h_L = \left(\frac{1}{C_c} - 1\right)^2 \frac{V^2}{2g}$$

C_c = coefficient of contraction
V = velocity in smaller-diameter pipe, ft/s

Bends and Standard Fitting Losses

The head loss that occurs in pipe fittings, such as valves and elbows, and at bends is given by

$$h_L = \frac{KV^2}{2g}$$

The table below gives some typical K values for these losses.

To obtain losses in bends other than 90°, the following formula may be used to adjust the K values:

$$K' = K\sqrt{\frac{\Delta}{90}}$$

Δ = deflection angle, deg

Fitting	K
Globe valve, fully open	10.0
Angle valve, fully open	5.0
Swing check valve, fully open	2.5
Gate valve, fully open Closed-return bend	0.2

Orifices

An orifice is an opening with a closed perimeter through which water flows. Orifices may have any shape, although they are usually round, square, or rectangular.

Orifice Discharge into Free Air

Discharge through a sharp-edged orifice may be calculated from

$$Q = Ca\sqrt{2gh}$$

Q = discharge, ft^3/s
C = coefficient of discharge
a = area of orifice, ft^2
g = acceleration due to gravity, ft/s^2
h = head on horizontal center line of orifice, ft

Submerged Orifices

Flow through a submerged orifice may be computed by applying Bernoulli's equation as follows:

$$V_2 = \sqrt{2g\left(h_1 - h_2 + \frac{V_1^2}{2g} - h_L\right)}$$

h_L = losses in head, ft.
Assuming $V_1 \approx 0$, setting $h_1 - h_2 = \Delta h$, and using a coefficient of discharge C to account for losses, one obtains:

$$Q = Ca\sqrt{2g\Delta h}$$

Discharge under Falling Head

The flow from a reservoir or vessel when the inflow is less than the outflow represents a condition of falling head. The time required for a certain quantity of water to flow from a reservoir can be calculated by equating the volume of water that flows through the orifice or pipe in time dt to the volume decrease in the reservoir:

$$Ca\sqrt{2gy}dt = Ady$$

Solving for dt yields

$$dt = \frac{Ady}{Ca\sqrt{2gy}}$$

a = area of orifice, ft^2

A = area of reservoir, ft^2

y = head on orifice at time t, ft
C = coefficient of discharge
g = acceleration due to gravity, $32.2\mathrm{ft/s}^2$

Expressing the area as a function of $y[A = F(y)]$ and summing from time zero, when $y = h_1$, to time t, when $y = h_2$, one obtains:

$$t = \int_{h_2}^{h_1} \frac{F(y)dy}{Ca\sqrt{2gy}}$$

If the area of the reservoir is constant as y varies, the above equation upon integration becomes

$$t = \frac{2A}{Ca\sqrt{2g}}\left(\sqrt{h_1} - \sqrt{h_2}\right)$$

h_1 = head at the start, ft
h_2 = head at the end, ft
t = time interval for head to fall from h_1 to h_2, s

<u>Fluid Jets</u>

Where the effect of air resistance is small, a fluid discharged through an orifice into the air will follow the path of a projectile. The initial velocity of the jet is

$$V_0 = C_v\sqrt{2gh}$$

where h = head on center line of orifice, ft

C_v = coefficient of velocity

The direction of the initial velocity depends on the orientation of the surface in which the orifice is located. For simplicity, the following equations were determined assuming the orifice is located in a vertical surface. The velocity of the jet in the X direction (horizontal) remains constant.

$$V_x = V_0 = C_v\sqrt{2gh}$$

The velocity in the Y direction is initially zero and thereafter a function of time and the acceleration of gravity:

$$V_y = gt$$

The X coordinate at time t is

$$X = V_x t = tC_v\sqrt{2gh}$$

The Y coordinate is

$$Y = V_{\text{avg}}t = \frac{gt^2}{2}$$

where V_{avg} = average velocity over period of time t. The equation for the path of the jet, is that for a parabola:

$$X^2 = C_v^2 4hY$$

$$C_v = \sqrt{\frac{X^2}{4hY}}$$

Pipe stresses

Pipe Stresses Perpendicular to the Longitudinal Axis

Internal pressure creates a stress commonly called hoop tension. The forces in the vertical direction cancel out. The sum of the forces in the horizontal direction is

$$pD = 2F$$

p = internal pressure, psi
D = outside diameter of pipe, in

F = force acting on each cut of edge of pipe, lb

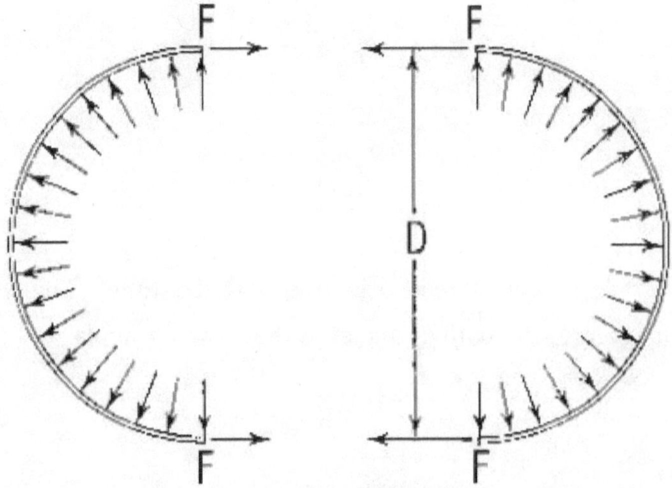

Hence, the stress, psi, on the pipe material is

$$f = \frac{F}{A} = \frac{pD}{2t}$$

A = area of cut edge of pipe, ft^2
t = thickness of pipe wall, in

<u>Temperature Expansion of Pipe</u>

The stress, psi, due to a temperature change is

$$f = cE\Delta T$$

E = modulus of elasticity of pipe material, psi
ΔT = temperature change from installation temperature
c = coefficient of thermal expansion of pipe material

The movement that should be allowed for, if expansion joints are to be used, is

$$\Delta L = Lc\Delta T$$

ΔL = movement in length L of pipe
L = length between expansion joints

Open channel flow

Basic Elements of Open Channels

The discharge Q at any section is defined as the volume of water passing that section per unit of time. It is expressed in cubic feet per second, ft^3/s, and is given by

$$Q = VA$$

V = average velocity, ft/s

A = cross-sectional area of flow, ft^2

When the discharge is constant, the flow is said to be continuous and therefore

$$Q = V_1A_1 = V_2A_2 = \cdots$$

where the subscripts designate different channel sections. The first equation above is known as the continuity equation for continuous steady flow.

Depth of flow d is taken as the vertical distance, ft , from the bottom of a channel to the water surface. The wetted perimeter is the length, ft , of a line bounding the cross-sectional area of flow, minus the free surface width. The hydraulic radius R equals the area of flow divided by its wetted perimeter. The average velocity of flow V is defined as the discharge divided by the area of flow,

$$V = \frac{Q}{A}$$

The velocity head H_V, ft, is generally given by

$$H_V = \frac{V^2}{2g}$$

V = average velocity, ft/s
g = acceleration due to gravity, 32.2ft/s^2

The true velocity head may be expressed as

$$H_{Va} = \alpha \frac{V^2}{2g}$$

α is an empirical coefficient that represents the degree of turbulence.

Experimental data indicate that α may vary from about 1.03 to 1.36 for prismatic channels. It is, however, normally taken as 1.00 for practical hydraulic work and is evaluated only for precise investigations of energy loss.

The total energy per pound of water relative to the bottom of the channel at a vertical section is called the specific energy head H_e. It is composed of the depth of flow at any point, plus the velocity head at the point. It is expressed in feet as

$$H_e = d + \frac{V^2}{2g}$$

Normal depth of flow

It may be computed by a trial-and-error process when the channel shape, slope, roughness, and discharge are known. A form of the Manning equation has been suggested for this calculation.

$$AR^{2/3} = \frac{Qn}{1.486S^{1/2}}$$

A = area of flow, ft^2
R = hydraulic radius, ft
Q = amount of flow or discharge, ft^3/s
n = Manning's roughness coefficient
S = slope of energy grade line or loss of head, ft , due to friction per lin ft of channel
$AR^{2/3}$ is referred to as a section factor.

Here is another interesting equation for open channel flow:

$$\frac{A^3}{T} = \frac{Q^2}{g}$$

in which T is the top width of the channel, ft . For a rectangular channel, the above equation reduces to

$$\frac{d}{2} = \frac{V^2}{2g}$$

$V = Q/A$ = mean velocity of flow, ft^3/s
d = depth of flow, ft

Critical Depth of Open-Channel Flow

Critical depth may be computed for a uniform channel once the discharge is known.

$$d_c = \sqrt[3]{\frac{Q^2}{b^2 g}}$$

d_c = critical depth, ft
Q = quantity of flow or discharge, ft^3/s
b = width of channel, ft

Manning's Equation for Open Channels

One of the more popular of the numerous equations developed for determination of flow in an open channel is Manning's variation of the Chezy formula,

$$V = C\sqrt{RS}$$

R = hydraulic radius, ft
V = mean velocity of flow, ft/s
S = slope of energy grade line or loss of head due to friction, ft/linft of channel
C = Chezy roughness coefficient

Manning proposed

$$C = \frac{1.486^{1/6}}{n}$$

where n is the coefficient of roughness in the earlier Ganguillet-Kutter formula. When Manning's C is used in the Chezy formula, the Manning equation for flow velocity in an open channel results:

$$V = \frac{1.486}{n} R^{2/3} S^{1/2}$$

Since the discharge $Q = VA$, one obtains:

$$Q = \frac{1.486}{n} A R^{2/3} S^{1/2}$$

A = area of flow, ft^2
Q = quantity of flow, ft^3/s

Values of the roughness coefficient *n* for use in the Manning equation are shown in the table below.

	Min	Avg	Max
A. Open-channel flow in closed conduits			
1. Corrugated-metal storm drain	0.021	0.024	0.030
2. Cement-mortar surface	0.011	0.013	0.015
3. Concrete (unfinished)			
a. Steel form	0.012	0.013	0.014
b. Smooth wood form	0.012	0.014	0.016
c. Rough wood form	0.015	0.017	0.020
B. Lined channels			
1. Metal			

a. Smooth steel (unpainted)	0.011	0.012	0.014
b. Corrugated	0.021	0.025	0.030
2. Wood			
a. Planed, untreated	0.010	0.012	0.014
3. Concrete			
a. Float finish	0.013	0.015	0.016
b. Gunite, good section	0.016	0.019	0.023
c. Gunite, wavy section	0.018	0.022	0.025
4. Masonry			
a. Cemented rubble	0.017	0.025	0.030
b. Dry rubble	0.023	0.032	0.035
5. Asphalt			
a. Smooth	0.013	0.013	
b. Rough	0.016	0.016	
C. Unlined channels			
1. Excavated earth, straight and uniform			
a. Clean, after weathering	0.018	0.022	0.025
b. With short grass, few weeds	0.022	0.027	0.033
c. Dense weeds, high as flow depth	0.050	0.080	0.120
d. Dense brush, high stage	0.080	0.100	0.140
2. Dredged earth			

a. No vegetation	0.025	0.028	0.033
b. Light brush on banks	0.035	0.050	0.060
3. Rock cuts			
a. Smooth and uniform	0.025	0.035	0.040
b. Jagged and irregular	0.035	0.040	0.050

Hydraulic jump

Depth and Head Loss in a Hydraulic Jump

According to Newton's second law of motion, the rate of loss of momentum at the jump must equal the unbalanced pressure force acting on the moving water and tending to retard its motion. This unbalanced force equals the difference between the hydrostatic forces corresponding to the depths before and after the jump. For rectangular channels, this resultant pressure force is

$$F = \frac{d_2^2 w}{2} - \frac{d_1^2 w}{2}$$

d_1 = depth before jump, ft

d_2 = depth after jump, ft

w = unit weight of water, lb/ft^3

The rate of change of momentum at the jump per foot width of channel equals

$$F = \frac{MV_1 - MV_2}{t} = \frac{qw}{g}(V_1 - V_2)$$

M = mass of water, $\text{lb} \cdot \text{s}^2/\text{ft}$
V_1 = velocity at depth d_1, ft/s
V_2 = velocity at depth d_2, ft/s
q = discharge per foot width of rectangular channel, ft^3/s
t = unit of time, s
g = acceleration due to gravity, 32.2ft/s^2

The reduced equation for rectangular channels becomes

$$V_1^2 = \frac{gd_2}{2d_1}(d_2 + d_1)$$

The above equation may then be solved for the sequent depth:

$$d_2 = \frac{-d_1}{2} + \sqrt{\frac{2V_1^2 d_1}{g} + \frac{d_1^2}{4}}$$

$$d_1 = \frac{-d_2}{2} + \sqrt{\frac{2V_2^2 d_2}{g} + \frac{d_2^2}{4}}$$

The head loss in a jump equals the difference in specific-energy head before and after the jump. This difference is given by

$$\Delta H_e = H_{e1} - H_{e2} = \frac{(d_2 - d_1)^3}{4d_1 d_2}$$

H_{e1} = specific-energy head of stream before jump, ft
H_{e2} = specific-energy head of stream after jump, ft

The depths before and after a hydraulic jump may be related to the critical depth by the equation

$$d_1 d_2 \frac{d_1 + d_2}{2} = \frac{q^2}{g} = d_c^3$$

q = discharge, ft^3/s per ft of channel width
d_c = critical depth for the channel, ft

<u>Jump in Horizontal Rectangular Channels</u>

Hydraulic depth is defined as

$$d_h = \frac{A}{T}$$

A = area of flow, ft^2
T = width of free surface, ft

For rectangular channels, hydraulic depth equals depth of flow.

Wells

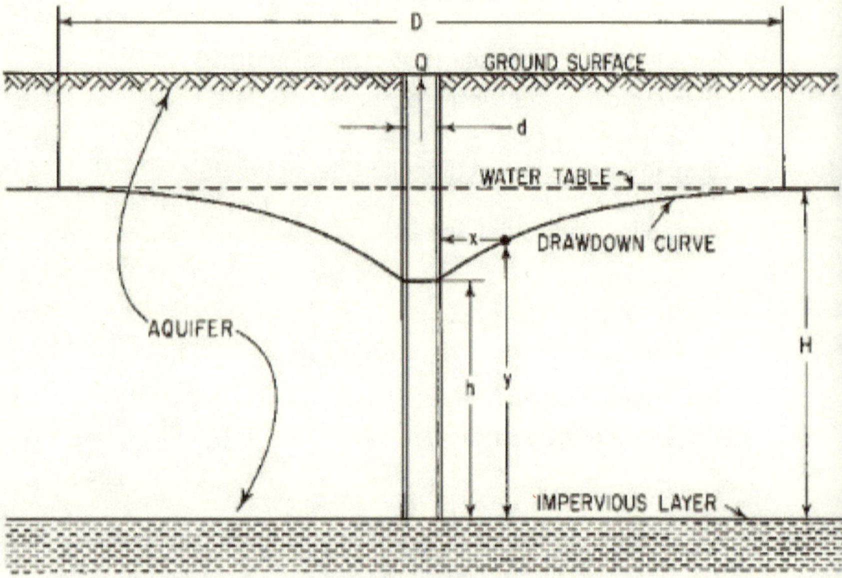
D
Q
GROUND SURFACE
d
WATER TABLE
x
DRAWDOWN CURVE
AQUIFER
H
h
y
IMPERVIOUS LAYER

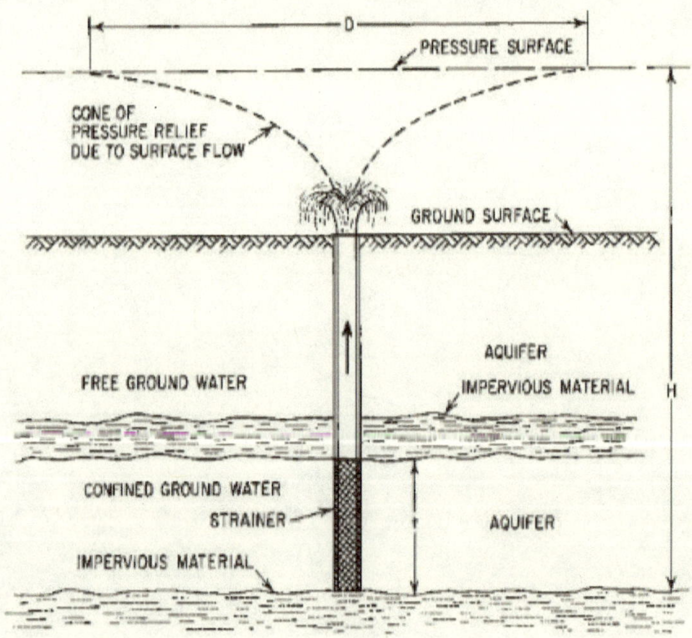
D
PRESSURE SURFACE
CONE OF
PRESSURE RELIEF
DUE TO SURFACE FLOW
GROUND SURFACE
AQUIFER
FREE GROUND WATER
IMPERVIOUS MATERIAL
H
CONFINED GROUND WATER
STRAINER
AQUIFER
IMPERVIOUS MATERIAL

Groundwater

Aquifers are groundwater formations capable of furnishing an economical water supply. Those formations from which extractions cannot be made economically are called aquicludes.

Permeability indicates the ease with which water moves through a soil and determines whether a groundwater formation is an aquifer or aquiclude.

The rate of movement of groundwater is given by Darcy's law:

$$Q = KIA$$

Q = flow rate, gal/day

K = hydraulic conductivity, ft/ day or m/ day

I = hydraulic gradient, ft/ft or m/m

A = cross-sectional area, perpendicular to direction of flow, ft^2 or m^2

Drawdown

When water is pumped from a well, the water level around the well draws down and forms a cone of depression – see the figure above. The line of intersection between the cone of depression and the original water surface is called the circle of influence.

Since nearly all soils are heterogeneous, pumping tests should be made in the field to determine the value of the hydraulic conductivity K. A permeability analysis of a soil sample that is not representative of the soil throughout the aquifer would produce an unreliable value for K.

Flow From Wells

The steady flow rate Q can be found for a gravity well by using the Dupuit formula:

$$Q = \frac{1.36K(H^2 - h^2)}{\log(D/d)}$$

Q = flow, gal/day

K = hydraulic conductivity, ft/ day under 1: 1 hydraulic gradient

H = total depth of water from bottom of well to free-water surface before pumping, ft

h = H minus drawdown, ft

D = diameter of circle of influence, ft

d = diameter of well, ft

The steady flow, gal/day, from an artesian well is given by

$$Q = \frac{2.73Kt(H - h)}{\log(D/d)}$$

Where t is the thickness of the confined acquifer.

Environmental Engineering

Sewer Design

Velocity Formulas

Velocity of flow, ft/s, in straight sewers without obstructions may be estimated with satisfactory accuracy from the Manning formula

$$V = \frac{C}{n} R^{2/3} S^{1/2}$$

n = coefficient dependent on roughness of conduit surface
R = hydraulic radius, ft = area, ft^2, of fluid divided by wetted perimeter, ft
S = energy loss, ft/ft of conduit length; approximately the slope of the conduit invert for uniform flow
$C = 1.486$ (conversion factor to account for change from metric units used in development of the formula)

A common value for n is 0.013 , suitable for well-laid brickwork, smooth concrete pipe, and vitrified clay pipe with liner plates. For vitrified clay pipes without liner plates, and plastic and resin-lined pipes, 0.011 may be used for n for design purposes. For corrugated-metal pipe, n may range from 0.011 for a spun asphalt lining to 0.02 for the plain pipe or pipe with a paved invert. Smaller values of n than the preceding may be used for new smooth pipe, but the roughness, and value of n, is likely to increase with age. The quantity of flow, ft^3/s, is given by

$$Q = AV$$

A = cross-sectional area of flow, ft^2.

The quantity, ft^3/s, for flow full may be estimated from

$$Q = \frac{0.463}{n} d^{8/3} S^{1/2}$$

and the velocity for flow full from

$$V = \frac{0.59}{n} d^{2/3} S^{1/2}$$

d = inside diameter of pipe, ft .

Storm-Water Inlets

An inlet may be a curb opening, a gutter grating, or a combination of the two. Capacity of the curb opening type when diverting 100% of gutter flow may be computed from

$$Q = 0.7\,\mathrm{L}(a + y)^{3/2}$$

Q = quantity of runoff, ft^3/s
L = length of opening, ft
a = depression in curb inlet, ft
y = depth of flow at inlet, ft

In practice, the gutter may be depressed up to 5 in below the normal gutter line along the length of the inlet. Slope of the gutter commonly is 1 in in 12 in . The depression may extend up to 3 ft from the curb. Depth of flow in the gutter may be estimated from the Manning formula.

For depths of flow up to 0.4 ft capacity of inlet may be calculated from the weir formula

$$Q = 3Py^{3/2}$$

P = perimeter, ft , of grate opening over which water may flow, ignoring the bars. For depths of flow greater than 1.4 ft , capacity may be computed from the orifice formula

$$Q = 0.6A\sqrt{2gy}$$

A = total area of clear opening, ft^2 g = acceleration due to gravity, $32\mathrm{ft/s}^2$

Side weirs

Diversion, ft^3/s, may be estimated from the Engels formula:

$$Q = 3.32l^{0.83}h^{1.67}$$

l = length of weir, ft
h = depth of flow over weir at downstream end, ft

Siphon Spillways

Area, ft^2, of the siphon throat can be determined from

$$A = \frac{Q}{c\sqrt{2gh}}$$

Q = discharge, ft^3/s
c = coefficient of discharge, which varies from 0.6 to 0.8
g = acceleration due to gravity $= 32.2\mathrm{ft/s}^2$
h = head, ft

Friction Loss in Full Pipe Flow

Friction loss in full pipe flow consists of losses from pipe friction, H_P as well as losses from pipe fittings, valves, and similar losses, generally referred to as minor losses, H_M. Pipe friction losses for full pipe flow can be calculated from the Hazen-Williams formula (adapted for circular pipe):

$$S = \left[\frac{V}{1.318C(D/4)^{0.63}}\right]^{1.8519}$$

S = energy loss, ft/ft of pipe length
V = velocity of flow, ft/s
C = coefficient of friction
D = pipe inside diameter, ft

Coefficient of friction, C, varies based on the type of pipe material and its age. Typical values for C are:

130 for new ductile iron pipe
100 for old ductile iron pipe
150 for new PVC pipe
140 for old PVC pipe.

Pipe friction loss, H_P, ft, can be determined for a given length of pipe L, ft, by:

$$H_P = SL$$

Minor losses, H_M, ft, are generally a function of the velocity head. The following equation is used to calculate minor losses:

$$H_M = K\frac{V^2}{2g}$$

K = minor loss coefficient
V = pipe velocity, ft/s

g = acceleration due to gravity = 32.2ft/s^2

Work and Pump Efficiency

The head against which a pump works is defined as pumping head, which is calculated as follows:

$$P_H = H_P + H_M + S$$

P_H = pumping head, ft
H_P, ft given in the previous section
H_M, ft given in the previous section
S = static head, ft

The power required to operate a centrifugal pump, P, is calculated as follows:

$$P = \frac{QP_H}{3960E_P}$$

P = power, HP
Q = flow rate, gpm
P_H = pump head, ft
E_P = pump efficiency, dimensionless

The efficiency of a pumping system, E_S, also known as wire to water efficiency, can be calculated as follows:

$$E_S = E_P E_M$$

E_S = system efficiency, dimensionless
E_P = pump efficiency, dimensionless
E_M = motor efficiency, dimensionless

Motor efficiencies are available from the motor manufacturers.

Characteristics of domestic wastewater

BOD, CBOD, and COD

BOD is determined by diluting a wastewater sample with water with known dissolved-oxygen content and storing the mixture for 5 days at 20°C. The oxygen content at the end of the period is measured, and the difference is reported as the BOD.

At the end of a period of t days at 20°C

$$\mathrm{BOD} = O(1 - 100^{-K_1 t})$$

O = oxygen demand when $t = 0$ or at start of any oxidation period

K_1 = deoxygenation coefficient, usually about 0.1 for wastewater, but may range from less than 0.05 to more than 0.2 . For temperatures other than 20°C, multiply K_1 for 20°C by 1.047^{T-20}. Regression analyses of laboratory data should be used to determine K_1 for the wastewater being tested.

T = temperature, °C.

To obtain the initial oxygen demand at temperatures other than 20°C, multiply O for 20°C by $0.02T + 0.6$.

References

Standard Handbook for Civil Engineers by Frederick S. Merritt, Third Edition, McGraw-Hill, 1983.

Standard Handbook for Civil Engineers by Jonathan Ricketts, M. Loftin, and Frederick S. Merritt, Fifth Edition, McGraw-Hill, 2004.

Standard Handbook of Engineering Calculations, Fifth Edition, by Tyler G. Hicks, McGraw-Hill, 2014.

Engineers' Practical Databook: A Technical Reference Guide for Students and Professionals, by Jay Smith, Independently published. 2018.

Mathematical Handbook for Scientists and Engineers: Definitions, Theorems, and Formulas for Reference and Review (Dover Civil and Mechanical Engineering), Revised Edition, by A. Korn and Theresa M. Korn, Dover Publications, 2000.

Structural Engineering Formulas, Second Edition, by MIKHELSON, McGraw-Hill, 2013.

Engineering Formulas 8th Edition, by Kurt Gieck and Reiner Gieck, McGraw-Hill, 2006.

www.ingramcontent.com/pod-product-compliance
Lightning Source LLC
LaVergne TN
LVHW090433210926
848480LV00002B/814
* 9 7 9 8 3 4 9 3 3 1 3 5 0 *